YOUR KNOWLEDGE HAS VALUE

- We will publish your bachelor's and
  master's thesis, essays and papers

- Your own eBook and book -
  sold worldwide in all relevant shops

- Earn money with each sale

Upload your text at www.GRIN.com
and publish for free

Economic efficiency in wheat production. Analysis of the case of Angecha District, Southern Ethiopia

Alemayehu Bashe
Asefa Ayele
Matewos Matselo
Tessema Erchafo
Seyfu T/yohannes

Bibliographic information published by the German National Library:

The German National Library lists this publication in the National Bibliography; detailed bibliographic data are available on the Internet at http://dnb.dnb.de.

ISBN: 9783346802002
This book is also available as an ebook.

Print and binding: Books on Demand GmbH, Norderstedt, Germany
Printed on acid-free paper from responsible sources.

The present work has been carefully prepared. Nevertheless, authors and publishers do not incur liability for the correctness of information, notes, links and advice as well as any printing errors.

GRIN web shop: https://www.grin.com/document/1320287

Analysis of Economic efficiency in wheat production: The case of Angecha District, southern Ethiopia

ABSTRACT

Production improvement through the use of improved technologies and increasing efficiency of inputs in cereal production in general and wheat production in particular might be an important alternative to settle food security problem in Ethiopia. Wheat is the first cash crop produced in Angecha District, kembata Tambaro zone, with a total area covered of 4567.5ha. But, the efficiency of producers that they could not use available resources on hand was taken as a great attention. So this study was aimed to analyze the levels of technical, allocative and economic efficiencies of wheat producers; and determining factors for inefficiency in farmers' wheat production by using cross sectional data from randomly selected 123 households in 2018/19 production year. The study used both primary and secondary data sources and stochastic production frontier approach was used to estimate the level of efficiencies. Ordinary least square estimation was used to identify factors that affect inefficiencies of sample farmers' in study area. The regression model result indicated that input variables like land and seed were the significant variables to increase the yield of wheat output. 55.63%, 55.47% and 30.85% were the estimated mean values of technical, allocative and economic efficiencies respectively, which indicate the presence of inefficiency in wheat production in the study area. Model result indicated that technical inefficiency positively and significantly affected by sex of the household head, and negatively affected by age, farm experience, land fragmentation, credit access and total livestock unit. Similarly, allocative inefficiency positively and significantly affected by sex and negatively by credit access and total livestock holdings. In addition, economic inefficiency negatively and significantly affected by credit access and total livestock holdings. The policy measures implied from the results include: working further for quality seed and sustainable land management, expansion of gender sensitive and youth based strengthening of the extension services and trainings, strengthening the existing credit institutions services, and expansion of new livestock technologies in the study area.

Key words: Angecha, economic efficiency, frontier, wheat production,

Table of Contents

1. INTRODUCTION

1.1. Background

In sub-Saharan Africa, Ethiopia is the second largest producer of wheat, following South Africa. It is a staple food in the diets of several Ethiopian, providing about 15% of the caloric intake (FAO, 2015), hence, increasing productivity in smallholder agriculture is government top priority, recognizing the importance of the smallholder sub-sector, the high prevalence of rural poverty and the large productivity gap. The country has a large potential for wheat production, which has yet to be fully realized due to various production-related issues. For example, Hei et al. (2017), Abate (2018), Tesfaye et al. (2018), and Ayele et al. (2019) found that a lack of improved varieties, poor seed supply systems, producers' reliance on local seeds, high fertilizer and seed costs, poor agronomic practices, weeds, pests, and diseases, weak farmers organizations, poor market information systems, and little research support to increase yields and climate change are major constraints to Ethiopian wheat production. Even though wheat production has increased significantly over the last two decades as a result of various government programs and efforts aimed at improving agricultural growth and food security in the country, domestic wheat grain and flour self-sufficiency remains a long way off (Gebreselassie et al., 2017). For example, the Ethiopian Grain Trade Enterprise imported 1.05 million tonnes of wheat in 2016 (Brasesco et al., 2019). On the other hand, Wheat output markets in Ethiopia are characterized by an insufficient transportation network, a small number of traders, insufficient capital facilities, high handling costs, an insufficient market information system, farmers' poor bargaining power, and underdeveloped industrial sectors (Mohammed and Addisu, 2016; Mamo et al., 2018). Domestic wheat consumption increased 41 percent, from 3.72 million tons in 2010 to 5.25 million tons in 2014, Mamo et al. (2018).

The agricultural sector in Ethiopia is explained by low productivity, caused by combination: of natural hazards, demographic factors, socio-economic factors; lack of knowledge on the efficient utilization of available and limited resources, poor and backward technologies and limited use of modern agricultural technologies (WFP, 2012). During the past years, the government and NGOs have undertaken various attempts to enhance agricultural productivity particularly that of cereal crops so as to achieve food security and to reduce poverty in the country.

Researchers from the International Food Policy Research Institute (IFPRI) collected wheat growing farmer survey to how wheat growers in Ethiopia respond to the new promotional package rolled out by the Ministry of Agriculture (MoA) and Ethiopian Agricultural Transformation Agency (ATA). The purpose of the package is to help wheat farmers increase their crop yields (IFPRI, 2015). There are about

4.7 million wheat farmers in Ethiopia. Of these, more than three-quarters (78 percent) live in Oromia and Amhara. In contrast, the smallest areas cultivated with wheat are found in SNNPR, where the average is just 0.19 ha/farm, which is dominated by small-scale farmers (Nicholas*etal*, 2015, Fetagn, 2018).

Economic efficiency is composed of two components; technical component and allocative component. The technical component refers to the ability to avoid waste, either by producing as much output as technology and input usage allow or by using as little input as required by technology and output production. And the allocative component refers to the ability to combine inputs and/or outputs in optimal proportions in light of prevailing prices (Fried *et al.*, 2008). Technical efficiency (that part of efficiency which explains the physical performance of a firm) measures the relative ability of a farmer to get the maximum possible output at a given input or set of inputs (Lovell, 1993). In SNNPR, the total area covered by wheat was 0.3 million hectares, which is produced from *0.53 million* smallholders' farmers with the total production of *3.39 million quintal*. The average productivity was 26.66qt/ha according to (CSA, 2017/18). According to Angecha District, Wheat is the first and major cereal crop with a total area of 4567.5hectar. Wheat is primarily produced as a cash crop food in the district. Wheat consumption as well as production currently increases alarmingly from time to time over the country Ethiopia and the District as well. Due to above mentioned reasons and others, it was aimed that there was no literature and past efficiency study conducted in the study area, but the area was well known in maximum production of wheat both at zonal and regional level. So that it was aimed to conduct economic efficiency in terms of productive and cost efficiency and to document the data as reference to scholars, stakeholders, governments and non-governmental institutions by extracting full of information through estimating the level of technical, allocative and economic efficiencies in wheat production and identifying the major determinants that lead variations in efficiencies in wheat producer households in the study area.

2. METHODOLOGY

2.1. Description of the study area

The study was conducted in Angecha District, Kambata Tambaro Zone of South nation nationalities and people's region of Ethiopia. Part of Kembata Tembaro Zone, Angecha is bordered on north by Hadya Zone, on west by Doyogena, on south by Kacha Bira, on the east by Damboya, and on the southeast by Kedida Gamela. The area of the District is mainly of 35% dega, 65% woina-dega and its altitude ranging from 1900-3018 meter above sea level. The area wa characterized with Minimum and maximum temperature of 12 and 16C^0. The District receives an average annual rain fall of 1250ml. Angecha has 77 km of all-weather roads and 45 of dry weather roads. The area practice mixed crop-livestock farming system. Wheat is the first major cereal crop followed by teff, faba bean, field pea, barley and sorghum.

2.2. Sampling technique and sample size determination

Two stages random sampling procedures was employed to draw a representative sample the District. In the first stage, two kebeles out of the 10 maximum wheat producing kebeles in the district were randomly selected. In the second stage, 123 sample farmers were selected using simple random sampling technique based on probability proportional to the size of wheat producers in each of two selected kebeles. To obtain a representative sample size, the study employed the sample size determination formula given by Yamane (1967) (Eq. 1).

$$n = \frac{N}{1+N(e^2)} \ldots \ldots \ldots \ldots \ldots \ldots \ldots \ldots \ldots \ldots . (1)$$

Where: n- Sample size; N- total number of wheat producing household heads in the District which account (4324); e- margin error (8%)

2.3. DATA COLLECTION, TYPE AND SOURCES

Qualitative and quantitative data were used in order to meet the objectives. Data was collected from both primary and secondary data sources. The primary data were obtained through structured questionnaire designed that was administered by the trained enumerators. The questionnaire was pre-tested and important corrections were made before direct use. Secondary data were also collected from bureau of agriculture of the district and other relevant sources.

2.4. METHODS OF DATA ANALYSIS

Both descriptive and econometric analyses were used to narrate the data. Descriptive statistics, like mean, minimum, maximum, standard deviations, frequency and percentage were used. Most empirical studies on efficiency in Ethiopia were analyzed using stochastic production frontier method (Solomon, 2014; Ahmed *et al*, 2015; Sisay *et al.*, 2015, *Asfaw et al., 2019*). This estimation procedure guarantees that the assumption of independent distribution of the inefficiency error term is not violated. The maximum likelihood estimation of the stochastic frontier model yields the estimate for beta (β), sigma squared ($\sigma2$) and gamma (γ), and are variance parameters; γ value measures the total deviation of observed output from frontier output. The study used the parameterization following Battese and Coelli (1995) and is given as, $\sigma^2 = \sigma_v^2 + \sigma_u^2$ and, $\gamma = \sigma_u^2/\sigma_v^2 + \sigma_u^2$ where the gamma lies between zero and one ($0 \leq \gamma \leq 1$). If the value is very close to zero, then the deviations are due to the result of random factors and/or if the value is close to 1, then the deviations are due to inefficiency factors from the frontier function. The main reason is that stochastic approach allows for the estimation of statistical noise such as measurement error and climate change which are beyond the control of the decision making unit and specified in the model as:

$$\ln(Yi) = \beta_0 + \sum_{i=1}^{n} \beta_i lnX_i + v_i - u_i, \qquad i = 1,2,3 \dots \dots \dots (2)$$

Where, i represents the number of sample households;

$\ln(Yi)$ Represents the natural logarithm of scalar yield of wheat from i^{th} farm households;

X_i Denotes the vector of input variables used by i^{th} farmers;

β_i Represents the vector of unknown parameter to be estimated;

vi intended to represent a random variation in output due to statistical factors which is beyond the control of farmers.

ui Permits to capture inefficiency effects in the production of wheat measured as the ratio of observed output to maximum feasible output of the ith farm.

$$TE = \frac{F(Xi;\beta).\exp(v_i - u_i)}{F(Xi;\beta).\exp(v_i)} = exp(-u_i) \dots \dots \dots \dots \dots \dots \dots \dots \dots (3)$$

Where, $F(Xi;\beta).\exp(v_i - u_i)$ is an observed output of (Y) and F (Xi; β). exp(vi) is the frontier output (Y*).

Specifically the dual cost frontier functional form can be represented as follows:

$$\ln[(C)i] = \alpha + \sum_{j=1}^{6} \beta_j lnC_{ji} + v_i + u_i \dots \dots \dots \dots \dots \dots \dots (4)$$

Where:

lnC_i Denotes the log of cost incurred for production by i^{th} farm.

C_{ji} Permits the vector of input prices and output of ith firm; α, β_j were the vector of unknown parameters to be estimated; v_i denotes random variables assumed to be independent and identically distributed random errors with zero mean and variance (σv^2) and

u_i Represents non-negative random term assumed to denote cost inefficiency.so that allocative efficiency is calculated as the inverse of cost efficiency. Economic efficiency is the product of technical and allocative efficiencies (EE=TE*AE).

3. RESULTS AND DISCUSSIONS

3.1. Descriptive statistics of variables used in production and cost functions

(Table-1) Descriptive summary of variables used in production and cost functions of wheat(N=123)

Variables	Unit	Min	Max	Mean	Std. Dev
Yield	Qt.	.25	52.00	10.05	8.85
Land	Ha	.1250	2.00	.52	.37
Seed	Kg	12.5	200.0	53.75	33.56
NPS	Kg	.0001	300.00	61.03	41.25
Urea	Kg	.00	300.00	74.39	64.78
Labor	Man-days	4	101	30.02	15.23
Oxen	Oxen-days	6.0	44.0	16.94	6.69
Total cost of production	Birr	909.50	12230.00	3950.70	1828.49
cost of seed	Birr	250.00	3000.00	798.28	504.113
cost of NPS	Birr	.0	3600.0	742.857	509.06
Cost of urea	Birr	112.00	3600.00	775.59	706.72
Cost of labor	Birr	80.0	3800.0	955.305	697.13
Cost of oxen	Birr	90	2050	591.11	265.69
Cost of land	Birr	7.80	480.00	87.66	77.010

From above (Table.1) the mean output of wheat obtained from sample household is 10.05qt/ha, which is the dependent variable in production function. The average land allocated to produce wheat from sample household during the period of survey was 0.52ha. Similarly sample household incurred total of 3950.70(ETB) on average to produce 10.05qt of wheat at farm level. Among the factors of production used, v the highest share of cost fall in labor, seed, urea and NPS fertilizer respectively of 955.35, 798.28, 775.59 and 742.86 ETB.

3.2. Socioeconomic and demographic characteristics of variables in efficiency model

(Table-2) Socioeconomic characteristics of continuous variables in sampled households

Variables	Min	Max	Mean	Std. Dev
Age of household head in (years)	21	70	44.76	12.41
Family size in (ME)	1.20	12.60	4.92	2.05
Education level of household head in (years of schooling)	0	13	5.72	3.92
Farming experience of household head in (years)	1	50	22.62	11.55
Total land owned in household head in (ha)	.25	3.0	.93	.56
Land fragmentation of household in (number)	.0	7.0	2.22	1.16
Livestock owned in (TLU)	0	19	7.03	3.53

Survey result, 2019

The average age of wheat producer households in the study area was 44.76 years with the average farm experience in production of wheat around 22.62 years. This helps the farming households solve their farm related problems like farm equipment's and other input delivery on time. In addition, the sampled wheat producer households had total land holding of 0.93ha on average.

(Table-3) Summary of dummy variables in used in the model

Variables	Description	Frequency	Percentage
Sex of household head	Male(0)	109	88.6
	Female	14	11.4
Access to extension service	Yes (1)	104	84.6
	No	19	15.4
Participation in off/non-farm activities	Yes (1)	31	25.2
	No	92	74.8
Credit utilization	Yes (1)	56	54.5
	No	67	45.5
Perception to fertility status of soil	Yes (fertile) (1)	109	88.6
	No (infertile)	14	11.4

Survey result,2019.

Thus a stochastic frontier model is preferable because of its ability of capturing measurement error and other statistical noise computing the shape and position of the production frontier. stochastic frontier production model proposed by Battese and Coelli (1995) in accordance with the original models for Aigner *et al.* (1977), and Meesuen and van den Broeck (1977) was be applied to cross-sectional data to analyze the efficiency. The stochastic frontier production function of Cobb-Douglas type is defined in logarithmic forms as:

$$ln(yi) = \beta 0 + \beta 1 ln(area) + \beta 2 ln(seed) + \beta 3 ln(NPS) + \beta 4 ln(Urea) + \beta 5 ln(labor)$$
$$+ \beta 6 ln(oxen) + \varepsilon i (vi - ui) - - - - - - - - - - - - -(5)$$

Where: ln natural logarithms, y represents the wheat output level of sample farmer, area, represents the land allocated for wheat operation in hectares, labor is the quantity of labor power used by the farmer in man-days (which includes weeding, chemical spray, ploughing, threshing, harvesting etc.), oxen is the quantity of oxen power used by the farmer in oxen-day, NPS is the quantity of NPS fertilizers used in kilogram, Urea, is also urea fertilizer applied/used for wheat production and Seed represents the quantity of wheat seed used in kilogram; β is production coefficient (unknown parameters) to be estimated. εi is an error term made up of two components; v_i is random error having zero mean, N $(0, \delta^2 v)$ which is associated with random factors such as measurement error in production and weather effect which are not control of the farmers and assumed to be independently and identically distributed as N $(0, \delta^2 v)$ with a random error that is independent of u_i, $u_i i$ is the non-negative efficiency measured relative to the stochastic frontier that is firm not attaining maximum efficiency of production and ranges between 0 and 1, which is also assumed to be independently and identically distributed as half-normal at zero mean or truncated half-normal at mean μ, N $[\mu, \delta^2 u]$ or according to Greene (1990) with two parameter gamma distributions. N represents the number of farmers included in the survey of the farm.

3.3. Hypothesis test

In this study, three hypotheses were tested. Accordingly, the functional form that can best fit to the data at hand was selected by testing the null hypothesis which states that the coefficients of all interaction terms and square specifications in the trans-log functional forms are equal to zero $(H0: \beta ij = 0)$ against alternative hypothesis $(H1: \beta ij \neq 0)$. This test was made based on the value of likelihood ratio (LR) statistics which could be computed from the log likelihood values of both the Cobb-Douglas and Trans log functional forms using Equ. 6.

$$\lambda = -2 [logL (H0) - logL (H1)] - - - - - - - - - - - - (6)$$

The computed λ value in the above equation 6 was compared with the upper 5% critical value of the $\chi 2$ at the degree of freedom equals to the difference between the number of independent variables used in both functional forms (in this case degree of freedom =10). Therefore, the λ value computed was 15.34 and this value is lower than the upper 5% critical value of $\chi 2$ at 10 degrees of freedom (18.31) (Table). As a result, the null hypothesis was accepted and the Cobb-Douglas functional form best fits the data. The next second test is to test the null hypothesis that the inefficiency component of the total error term is equal to zero ($\gamma = 0$) compared with alternative hypothesis that inefficiency component different from zero. Therefore, the log likelihood ratio is calculated and compared with the $\chi 2$ value at a degree of freedom equal to the number of restrictions /the inefficiency component estimated by the full frontier, which is one (1) in this case for all models. As explained in table below, one-sided generalized λ test of $\gamma = 0$ provide a statistics of 11.2 for wheat production; which is significantly higher than the critical value of $\chi 2$ for the upper 5% at one degree of freedom (3.84).

Due to the result, the null hypothesis that stated wheat producers in the study area are fully efficient was rejected. The third hypothesis test was that all coefficients of the inefficiency in the model are likely equal to zero (i.e.H0: $\delta 0 = \delta 1 = \delta 2 = \cdots \delta 12 = 0$) against the alternative hypothesis, that states all parameter coefficients of the inefficiency effect model are not simultaneously equal to zero. This is also tested in the same way by calculating the λ value using the value of the log likelihood function under the stochastic frontier model (without explanatory variables of inefficiency effects, H0) and the full frontier model (with variables that are supposed to determine efficiency level of each farmer, H1). Using the formula in Equation (6), the value λ obtained was 54.74, which is higher than the critical $\chi 2$ value (15.51) at the degree of freedom equal to the number of restrictions to be zero (in this case the number of coefficients of the inefficiency effect model was 8). As a result, the null hypothesis is rejected in favor of the alternative hypothesis that explanatory variables associated in inefficiency model are simultaneously different from zero. Table-4. Generalized Likelihood Ratio test for hypothesis parameters of SPF

Null hypothesis Decision	df	λ	Critical value	
Ho: $\beta ij = 0$	10	15.34	18.31	Accept H0
Ho: $\gamma = 0$ H0	1	11.2	3.84	Not accept
Ho: $\delta 0 = \delta 1 = \delta 2 = ..\delta 12 = 0$ H0	8	54.74	15.51	Not accept

3.4. Parametric stochastic production frontier of MLE

(Table-5)Estimates of the Cobb Douglas frontier production function

Variables	Parameters	MLE Coef.	Std. Err.
Intercept	β_0	6.615648	.6732504
Lnarea	β_1	.5784695***	.1164422
Lnseed	β_2	.3352947**	.1319122
LnNPS	β_3	-.1673022	.1533594
Lnurea	β_4	.1797981	.1184581
Lnlabor	β_5	.0787196	.1222992
Lnoxn	β_6	-.1088914	.0772435
Variance parameters			
$(\sigma^{2)} = \sigma^2{}_u + \sigma^2{}_v$			.185
$(\lambda) = \sigma_u/\sigma_v$			1.592
Gamma (γ)			0.7170

Survey result, 2019

As Table-5 represents, result of the model showed that two of the input variables in the production function: land and seed had a positive and significant effect on the level of wheat production. Hence, increase in these inputs would increase production of wheat significantly as expected. As one percent increase in the size of land and amount of seed might increase wheat production by .579%, .335% respectively. In addition gamma value (γ) is 0.7170, which implies 71.70% of total variation in wheat output is due to technical inefficiency.

Table-6: Summary of level of efficiencies in sample households

Efficiencies	Min	Max	Mean	Std. Dev
TE	.06	.88	.5563	.19008
AE	.06	.97	.5547	.21017
EE	.01	.61	.3085	.14561

Survey result, 2019

Efficiency score and its distribution: The mean TE was found to be 55.63%. This indicates that in the short run farmers on average would decrease inputs (land, seed, labour and inorganic fertilizers) by 44.37% if they have to be technically efficient. In other words, it refers that if resources were efficiently utilized, the sampled farmers could increase current output by 44.37% using existing resources and level of technology. In the same way, the average allocative efficiency of wheat producers in the study area was 55.47%. It indicated that wheat producing farmers would save 44.53% of their current cost of inputs in a way to minimize cost. In addition, the mean economic efficiency of 30.85% prevails that an economically efficient farmer can produce 69.15% additional wheat shown in the (Table-5) above.

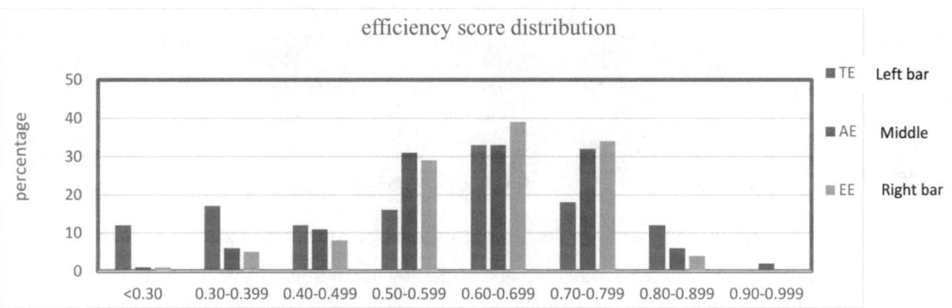

Figure.1: Frequency of efficiency score distribution of Technical, Allocative and Economic efficiencies

3.5. OLS estimation of different inefficiency variables

(Table-7). OLS regression result estimation for technical, allocative and economic inefficiencies

Variables	TE Coef.	TE Std. Err.	AE Coef	AE Std. Err	EE Coef	EE Std. Err
Age of household	-.004063*	.0020978	.0006638	.0013436	.0002916	.0012619
Sex of household head	.128714**	.0530219	.0681875**	.0339604	.0554291	.031894
Family size (ME)	.009086	.0086942	-.0035043	.0047981	-.0065143	.0052297
Education level of household head	-.000856	.0057846	.0111668	.0025677	.0005176	.0027634
Farming experience	-.005813**	.0023593	-.0075072	.0013103	-.007264	.0012783
Total land owned in (ha)	-.006685	.0310032	-.0380816	.0173751	-.0150723	.0163682
Land fragmentation	-.032750**	.014607	-.0114881	.0068958	-.0111394	.0069716
Credit access	-.069504**	.0354256	-.033648***	.0172559	-.0550004**	.0213105
Extension service	.002593	.0152341	-.0644314	.02269	.0425573	.0226487
Parti off/nonfarm	.037760	.030953	-.0258256	.0197869	-.0250219	.0198689
fertility status of soil	-.023238	.0552517	.1378887	.0308779	.0952941	.0308471
Total livestock in (TLU)	-.005055	.0043897	-.0062947*	.0032329	-.0069925**	.0030369

3.6. Summary of important explanatory variables in the inefficiencies

The age of household has a significant negative effect on inefficiency of wheat producing farmers at 10% significant level. This indicates that as the age of farmers' increases the inefficiency level decline which in turn increases technical efficiency of wheat producer farmers with more years of experience are expected to be less inefficient. The same result is obtained Endalkachew (2012) and Shumet (2012) and in contradiction with the study by Mesay *et al.* (2013), Assefa (2016). Moreover, farmers at older age might accumulate necessary resource control like oxen, farm tools and labour that could boost their efficiency, since in crop production, better availability of farm resources enhances timely application of inputs that increase efficiency of the farmer (Kitila and Alemu, 2014). Sex of household has significant and positive effect on TE and AE inefficiencies. Positive and significant effect indicates that inefficiencies increases in male household than females' counterparts. This might due to established settled female farmers' cooperative union on wheat seed and supply their production to other male headed farmers and other local traders and decide to easily access different extension services, trainings, mechanization services from other organizations and use new improved varieties which enhance productivity. Farming experience: the coefficient of farming experience of farm household on wheat production negatively affects the TE inefficiencies of farmers at 5% significant level. Its negative sign might indicate that those farmers having high experiences of farming were less inefficient and responsive for modern inputs combination from their long term farm proximity and maximize output. There is ease access to them experience input combination and there by maximizing output from them.

Scattered and fragmented land held by farmers was believed to affect inefficiency negatively, Farmers with more fragmented land (measured by the number of plots) are likely to be less inefficient than a farmer who has a few number of plots owned. The logical reasoning is that as the number of plots operated by a farmer increases, it becomes farm household only participate on agriculture sector there to focus on production potential and put himself as model and try to manage each plot more efficiently. In addition, a farmer who owned more of farm land may tries to use improved technologies and invest more of his/her resources for production improvement.

Credit access has significant and negative effect on all types of inefficiencies of smallholder wheat farmers in the study area. This refers that inefficiencies decrease in credit user households than non-users. The reason could be credit user farmers might spend and use it for wheat production timely and planned way in order for asset accumulation spending to increase the output of wheat production in cost minimizing way. Livestock owned (Lives): This is the total livestock holding in terms of Tropical Livestock Unit (TLU). Livestock would support crop production in different ways; they can be source of cash, draft power and manure that will be used to maintain soil fertility. It also serves as shock absorber to an unexpected hazard in crop failure and the main sources of animal labor in crop production. In addition, it indicates the wealth status of household. The number and value of livestock holding was found to be negatively related to technical inefficiency (Fekadu, 2004; Aynalem, 2006; Assefa, 2016) and also other inefficiencies. Due to this, in this study the effect of livestock on efficiency was hypothesized to be positive.

4. CONCLUSION AND RECOMMENDATION

The Cobb-Douglas approach of stochastic production frontier and its dual cost functions were applied from which TE, AE and EE extracted. The results of production function (TE) showed that, two of the factors of production (area and seed) were positively and significantly affect wheat output. In addition results showed that the input variables specified in the model had elastic effect on the output of wheat production. The coefficient calculated was 2.053, indicating increasing returns to scale. This indicates that there is potential for wheat farmers to expand their production due to they are in the stage I production area. This shows that, a 1 percent increase in all inputs proportionally might increase the total production of wheat by 2.053%. Therefore, an increase in all inputs by 1 percent could increase wheat output by more than 1percent. The average estimated level of TE, AE and EE of sampled farm households were 55.63%, 55.47% and 30.85% respectively.

This in turn implies that farmers can increase their wheat production on average by 36.78% when they were technically efficient. Similarly, they can reduce their cost by 42.78% given the optimum level of output. Furthermore it implies that the good resource utilizing base, improved efficiency can still be achieved and there exist a potential to increase the gross output and profit with the existing level of factor inputs. Using best practices of the efficient farmers as a point of reference would help setting targets in improving efficiency levels and focusing on the weakness of the current farm practices. In due the efficient farms might also improve their efficiency more likely through learning the best resource allocation decision from other technology demonstration factors. These could be achieved through field days, cross-visits, creating access for experience sharing with long period farm households and on job trainings. It shows that there is a base for wheat producers to increase wheat output at existing levels of inputs and minimize cost without negotiating yield with present technologies available in the hands of producers. Again; Among 12 explanatory variables hypothesized to determine inefficiencies; age, sex, farm experience, land fragmentation and credit access were found to be statistically significant to affect the level of technical inefficiency, sex of household head, access to credit service, Total livestock unit (TLU) affect allocative inefficiencies and access to credit services and Total livestock unit significantly affect the economic inefficiencies of sample wheat producer households in the study area. Therefore, the policy measures derived from the results include: expansion of gender sensitive and youth based strengthening of the extension services and trainings, establish and/or strengthening the existing credit institutions, developing and enhancing land management system and expansion of new livestock technologies in the study area.

5. REFERENCES

Aigner, et al (1977). Formulation and estimation of stochastic production function models. *Journal of Economics.* 6(1):21-37. DOI: https://doi.org/10.1016/0304-4076 (77)90052-5

Ahmed, B., Haji, J. and Geta, E. (2013). Analysis of Farm Households' Technical Efficiency in Production of Smallholder Farmers: The Case of Girawa District, Ethiopia. *American-Eurasian J. Agric. & Environ. Sci., 13 (12): 1615-1621.* DOI: 10.5829/idosi.aejaes.2013.13.12.12310

Ahmed, M. H., Lemma, Z. and Geta, E. (2015). Measuring Technical, Allocative and Economic Efficiency of Maize Production in Subsistence Farming: Evidence from the Central Rift Valley of Ethiopia. Applied Studies in Agribusiness and Commerce. DOI: 10.19041/Apstract/2015/3/9

Analysis of levels and determinants of technical efficiency of wheat producing farmers in Ethiopia. *African Journal of Agricultural Research. Vol. 11(36), Pp. 3391-3403.* https://doi.org/10.5897/AJAR2016.11310

Asfaw et al., (2019*)* Economic efficiency of smallholder farmers in wheat production: the case of Abuna Gindeberet District, Western Ethiopia. Review of Agricultural and Applied Economics https:// doi: 10.15414/raae.2019.22.01.65-75

Assefa, A. (2016). Technical Efficiency of Smallholder Wheat Production in Soro District of Hadiya Zone, Southern Ethiopia. MSc. Thesis Haramaya University.

Ayele, A., Erchafo, T., Bashe, A., and Tesfeyohannes, S., 2021. Value chain analysis of wheat in Duna District, Hadya Zone, southern Ethiopia. Heliyon (2021): e07597

Battese, G.E. and T.S. Coelli. 1995. Model for Technical Efficiency Effects in a stochastic frontier Production function for panel data. *Empirical Economics,* 20: 325-332.

Beshir H, Emana B, Kassa B, Haji J (2012). Economic efficiency of mixed crop-livestock production system in the North eastern highlands of Ethiopia: the stochastic frontier approach. J. Agric. Econ. Dev. 1(1):10-20.

CSA (Central Statistical Authority). 2018. Agricultural Sample Survey 2017/2018: Volume I Report on Area and Production of Major Crops (Private Peasant Holdings, *Maher* Season. Statistical Bulletin. Addis Ababa.

Debebe S, Haji J, Goshu D, Edriss KA (2015). Technical, allocative, and economic efficiency among smallholder maize farmers in Southwestern Ethiopia: Parametric approach. J. Dev. Agric. Econ. 7(8):282-291.

FAO (Food and Agriculture Organization) (2015). Food Balance Sheets. FAOSTAT. Rome.

Fekadu, G. and Bezabih, E. (2008). Analysis of Technical Efficiency of Wheat Production: A Study in Machakel Woreda, Ethiopia. MSc. Thesis Presented to the School of Graduate Studies, Haramaya University.

Fetagn G. (2018) Allocative Efficiency of Smallholder Wheat Producers in Damot Gale District, Southern Ethiopia

Geta, E., Bogale, A., Kassa, B., and Elias, E. (2013). Productivity and efficiency analysis of smallholder maize producers in Southern Ethiopia. *Journal of Human Ecology.* 41(1):67-75. DOI: https://doi.org/10.1080/09709274.2013.11906554

Kitila, G.M. and Alemu, B. A. (2014). Analysis of Technical Efficiency of Small Holder Maize Growing Farmers of Horo Guduru Wollega Zone, Ethiopia: A Stochastic Frontier Approach. *Science, Technology and Arts Research Journal,* Vol 3, No 3 (2014) pp 204-212. DOI: http://dx.doi.org/10.4314/star.v3i3.33

Mustefa B, Mulugeta T, Raja K. P, (2017). Economic efficiency in maize production in Ilu Ababor zone, Ethiopia

Solomon B. (2012). Economic efficiency of wheat seed production: the case of smallholders in Womberma Woreda of West Gojjam Zone. MSc. thesis presented to the School of Graduate Studies of Haramaya University.

Tiruneh, W. G. and Geta, E. (2016). Technical Efficiency of Smallholder Wheat Farmers: The Case of Welmera District, Central Oromia, Ethiopia. *Journal of Development and Agricultural Economics.* Vol. 8(2), pp.39-51. DOI: 10.5897/JDAE2015.0660

WFP (Food and Agricultural Organization and World Food Programme) (2012). Crop and Food Security Assessment Mission to Ethiopia. Special Report of Food and Agriculture Organization and World Food Programme.

Yamane, T. I. (1967). Statistics: An Introductory Analysis 2nd Edition. New York, Harper and Row.